1 MONTH OF
FREE
READING

at

www.ForgottenBooks.com

By purchasing this book you are eligible for one month membership to ForgottenBooks.com, giving you unlimited access to our entire collection of over 1,000,000 titles via our web site and mobile apps.

To claim your free month visit:
www.forgottenbooks.com/free1297126

ISBN 978-0-428-98071-9
PIBN 11297126

For support please visit www.forgottenbooks.com

Historic, archived document

Do not assume content reflects current
scientific knowledge, policies, or practices.

CONTRIBUTIONS TOWARD A

FLORA OF NEVADA.   NO. 34

X CUSCUTA OF NEVADA

by

T. G. YUNCKER

June 11, 1954

A series prepared through the cooperation of
the National Arboretum and the Section of Plant Introduction
Horticultural Crops Research Branch
Agricultural Research Service
U. S. Department of Agriculture
Plant Industry Station
Beltsville, Md.

Address all inquiries concerning this series
to W. Andrew Archer, Plant Industry Station,
Beltsville, Maryland

FOREWARD

The "Contributions Toward A Flora of Nevada" were begun in 1940
as an adjunct to the study of the medicinal uses of plants by the
Indian tribes of the State. The assembling of information from
the Indians and the collection of herbarium specimens were conducted
under a collaborating project between the Bureau of Plant Industry,
U. S. Department of Agriculture; the University of Nevada; and the
Work Projects Administration of Nevada.

The field work began in 1937 and continued until 1940. In the first
year a staff of some 20 collectors covered the entire State in an
attempt to secure a complete representation of the plants of Nevada.
The major portion of the collecting was done by Percy Train.
Approximately 14,000 numbers were accumulated during this time.

In the period 1940 to 1942 the following titles were released:

No.

1.	Swallen, J. R.	– Gramineae
2.	Muller, C. H.	– Fagaceae
3.	" " " "	– Aceraceae
4.	" " " "	– Betulaceae
5.	Fosberg, F. R.	– Santalaceae
6.	" " " "	– Ulmaceae
7.	McVaugh, Rogers	– Hypericaceae
8.	Fosberg, F. R.	– Chenopodiaceae
9.	Muller, C. H.	– Fouquieriaceae
10.	" " " "	– Vitaceae
11.	Hermann, F. J.	– Hydrocharitaceae
12.	" " " "	– Typhaceae

<u>No</u>.

13. Hermann, F. J.   – Sparganiaceae
14.   "   "   "   "   – Juncaginaceae
15.   "   "   "   "   – Naiadaceae
16.   "   "   "   "   – Juncaceae
17.   "   "   "   "   – Cyperaceae
18. Blake, S. F.   – Polygalaceae
19. Freeman, O. M.   – Lentibulariaceae
20.   "   "   "   "   – Menthaceae
21.   "   "   "   "   – Saururaceae
22. McVaugh, Rogers   – Rosaceae
23. Fosberg, F. R.   – Loganiaceae
24.   "   "   "   "   – Aizoaceae
25.   "   "   "   "   – Haloragaceae
26.   "   "   "   "   – Elaeagnaceae
27. McVaugh, Rogers   – Loasaceae
28. Muller, C. H.   – Amaryllidaceae
29. McVaugh, Rogers, & F. R. Fosberg – Index to the geographical names of Nevada
30. Erlanson, C. O.   – Violaceae
31. Martin, Robert F.   – Papaveraceae
32. Munz, Philip A.   – Onagraceae
33. Train, Percy, J. R. Henrichs & W. Andrew Archer – Medicinal uses of plants by Indian tribes of Nevada.

The series was interrupted because of the war, but it has now been decided to resume the work. Additional families will be released as the treatments are completed by various collaborators. Eventually all the material will be brought together and issued in book form.

It is to be noted that all the numbers are still available for distribution, except No. 33.

Address all inquiries concerning this series
to W. Andrew Archer, Plant Industry Station
Beltsville, Maryland

CUSCUTA OF NEVADA

By T. G. Yuncker*

CUSCUTA Linnaeus

CUSCUTA Linnaeus, Sp. Pl. 124, 1753.

Leafless and rootless, herbaceous parasites with yellowish filiform stems which twine about woody or herbaceous host plants from which
they obtain nourishment by means of haustoria; flowers small (mostly
2 to 6 mm. long), sessile or pedicellate, in few to many-flowered cy-
mose clusters, commonly 5-merous (a few species are regularly 3-4 merous);
perianth parts commonly more or less united; stamens inserted in the
throat of the corolla and alternating with the lobes; scale-like, more
or less toothed, fringed, or fimbriate appendages present in most
species at the base of the corolla opposite the stamens; ovary two-
celled, styles two; stigmas of various shapes (those of American species
are capitate); fruit a capsule which remains closed, or opens with a
regular or irregular line of circumscission near its base; embryo acoty-
ledonous, filiform or globose with a filiform appendage.

A genus with about 160 species of world-wide distribution.

KEY TO SPECIES

1. Stigmas capitate; capsules not easily or regularly opening near the
   base (American species.)
   2. Flowers mostly 3- or 4- parted, corolla lobes obtuse, the withered
      corolla remaining at the top of the capsule 1. C. cephalanthi

with inflexed tips - - - - - - - - - 2. C. <u>indecora</u>

4. Perianth not obviously fleshy-papillate; corolla lobes
   not entirely as above.

   5. Infrastamineal scales prominent, commonly exserted;
      corolla lobes triangular to sublanceolate, acute,
      reflexed; capsules mostly depressed-globose, 2-4
      seeded - - - - - - - - - - - - - - - - 3. C. <u>campestris</u>

   5. Scales included; capsules globose-conic, mostly 1-seeded.

      6. Flowers about 2 mm. long, subsessile; calyx lobes
         orbicular, broadly overlapping, margins of
         perianth lobes deticulate - - - - - 4. C. <u>denticu-</u>
         <u>lata</u>

      6. Flowers 2-4 mm. long, pedicellate; calyx lobes
         ovate-lanceolate or lanceolate.

         7. Corolla lobes ovate-lanceolate; scales
            attached to corolla tube most of their length;
            anthers oval, filaments well developed - - -
            - - - - - - - - - - - - - - - - 5. C. <u>salina</u>

         7. Corolla lobes lanceolate; scales commonly free;
            anthers oval-oblong, subsessile - - - - -
            - - - - - - - - - - - - - - - 6. C. <u>nevadensis</u>

3. Infrastamineal scales lacking.

   4. Flowers pedicellate; anthers oblong; capsules ovoid-
      pointed - - - - - - - - - - - 7. C. <u>californica</u> var. <u>apiculata</u>

   4. Flowers essentially sessile; anthers rounded or oval; cap-
      sules globose - - - - - - - - - - - - 8. C. <u>occidentalis</u>

1. Stigmas filiform, capsules easily opening near the base in a regular
   line of cleavage (introduced European species.)
   2. Calyx lobes triangular-lanceolate, apex not obviously thick and
      fleshy - - - - - - - - - - - - - - - - - 9. C. epithymum
   2. Calyx shallowly divided, lobes commonly wider than long, apex
      thick and turgid - - - - - - - - - 10. C. approximata var.
      urceolata

1. CUSCUTA CEPHALANTHI Engelmann, Amer. Jour. Sci. & Arts 43: 336. 1842

   Flowers up to about 2 mm. long from base to corolla sinuses,
   sessile or subsessile in open, paniculate-cymose clusters, mostly
   4-parted (less commonly 3- or 5-parted). Calyx shorter than the
   corolla tube, lobes oval- or oblong-ovate, obtuse. Corolla cylin-
   dric-campanulate, enlarging toward the base about the maturing capsule, lobes shorter than the tube, ovate, obtuse, erect to spreading.
   Infrastamineal scales oblong, about reaching the stamens, fringed
   with scattered processes. Stamens nearly equaling the corolla lobes;
   filaments stoutish and about as long as the oval to rounded anthers.
   Styles slender; stigmas globose-capitate. Capsule mostly nearly globose, sometimes asymmetrical because of irregularity of seed for-
   mation, capped by the withered corolla.

   Across the United States from Massachusetts to Washington and
   southward to Mexico, but most frequent eastward. This species
   occurs on many different species of woody and herbaceous hosts and
   exhibits little, if any, host preference.

   Nevada: Elko County.

2. CUSCUTA INDECORA Choisy, Mem. Soc. Phys. Hist. Nat. Geneve 9: 278

      pl. 3, f. 3. 1841

      Cuscuta neuropetala Engelmann, Amer. Jour. Sci. & Arts 45:

      75. 1843.

      Cuscuta pulcherrima Scheele, Linnaea 21: 750. 1849.

      Cuscuta decora Engelmann, Trans. Acad. Sci. St. Louis 1:

      501. 1859.

Flowers commonly 2-4 mm. long from the base to the corolla
sinuses, whitish, fleshy, papillose, on pedicels shorter than to
longer than the flowers, in compact or loose clusters. Calyx
shorter than or equaling the corolla tube, lobes triangular ovate,
acute or obtusish. Corolla campanulate, lobes mostly shorter than
the tube, erect to spreading, triangular, with acute, inflexed
tips. Infrastamineal scales equaling the corolla tube or slightly
exserted, rounded or somewhat spatulate, strongly fringed. Anthers
broadly oval, about equaling the filaments. Styles as long as or
slightly longer than the globose, pointed ovary; stigmas capitate.
Capsule globose, thickened about the base of the styles, enveloped
by the withered corolla.

    An attractive and wide-spread species parasitizing a great
variety of herbaceous and woody hosts.

    Nevada: Washoe County, on cultivated Aster.

3. CUSCUTA CAMPESTRIS Yuncker, Mem. Torr. Bot. Club 18: 138. 1932

      Cuscuta pentagona Engelmann var. calycina Engelmann, Amer.

      Jour. Sci. & Arts 45: 76. 1845.

Cuscuta __arvensis__ Beyrich var. calycina Engelmann, Trans.

Acad. Sci. St. Louis 1: 495. 1859

Flowers up to 2.5 mm. long from base to corolla sinuses, on pedicels mostly shorter than the flowers, in loose to compact, globular clusters. Calyx about as long as the corolla tube, lobes ovate to oval-ovate, obtuse. commonly nearly as long as wide, usually somewhat overlapping at the base but not obviously angled at the sinuses. Corolla lobes triangular to lanceolate, about as long as the campanulate tube. spreading to reflexed, tips acute and commonly inflexed. Infrastamineal scales abundantly fringed, exserted. Filaments longer than or about equaling the oval anthers. Styles slender, sometimes slightly thicker towards the base; stigmas capitate. Capsules commonly depressed-globose, with the withered corolla about the lower half.

This species occurs on a variety of hosts but exhibits a preference for various legumes, especially Trifolium, Medicago. etc., to which it often does considerable damage. A native of the United States, it has been distributed throughout the world presumably as a contaminant of the seeds of its leguminous hosts.

Nevada: Lincoln and Washoe Counties

4. CUSCUTA DENTICULATA Engelmann, Amer. Nat. 9: 348. 1875.

Flowers 1 to 1.5 mm. long from base to corolla sinuses, subsessile in the axils of lanceolate, denticulate, acute and somewhat squarrose bracts, solitary or in few-flowered glomerules. Calyx

enclosing the corolla tube, deeply divided,cells conspicuous, yellow
and glistening when dry, lobes suborbicular, overlapping, obtuse,
edges thin and denticulate. Corolla campanulate but soon becoming
urceolate about the maturing capsule, lobes oval-ovate, obtuse,
spreading, about as long as the tube, edges irregularly denticulate.
Infrastamineal scales reaching the stamens, denticulate. Stamens
shorter than the corolla lobes. Styles shorter than the small,
conic ovary; stigmas small, globose. Capsule conic, mostly 1-seeded
and bearing the withered corolla about the upper part; embryo globose with a short, curved, terete, tail-like appendage.

A species of the arid regions of Utah, Arizona, Nevada and California where it occurs on a number of different hosts.

Nevada: Clark, Douglas, Esmeralda, Lyon and Washoe Counties.

5. CUSCUTA SALINA Engelmann in Brewer, Watson & Gray, Bot. Calif.

1: 536. 1876.

Cuscuta californica var. squamigera Engelmann, Trans. Acad. Sci.

St. Louis 1: 499. 1859.

Cuscuta squamigera (Engelmann) Piper, Contrib. U. S. Nat. Herb.

11: 455. 1906.

Flowers about 2.5 mm. long from the base to the corolla sinuses,
on pedicels mostly shorter than the flowers, in umbellate- or paniculate-cymose clusters. Calyx lobes ovate-lanceolate, acute to acuminate, about as long as the corolla tube, often somewhat shiny-yellow
when dry. Corolla lobes about as long as the narrowly campanulate
or somewhat cylindrical tube, ovate-lanceolate, usually granulate,

acute to acuminate, upright to spreading. Infrastamineal scales narrow, oblong, commonly shorter than the tube, fringed with short processes, closely attached to the tube to near the top. Filaments subulate and about equal to or shorter than the oval anthers. Styles slender or slightly subulate, about equaling the ovary; stigmas capitate. Capsule globose-ovoid, slightly pointed, usually 1-seeded, surrounded or capped by the withered corolla.

From British Columbia to California and eastward to Utah, Nevada and Arizona. Commonly on halophytic hosts.

Nevada: Nye County.

6. CUSCUTA NEVADENSIS I. M. Johnston, Proc. Calif. Acad. Sci. IV. 12: 1133. 1924

Cuscuta salina Engelmann var. apoda Yuncker, Mem. Torr. Bot. Club 18: 169. 1932.

Flowers mostly about 2 mm. long from base to corolla sinuses, on pedicels up to as long as or sometimes exceeding the length of the flowers, solitary or in 2- or 3-flowered umbel-like clusters. Calyx lobes as long as or sometimes much exceeding the corolla tube, lanceolate, acute to acuminate, somewhat overlapping at the base, cells large and glistening yellow when dry. Corolla narrowly campanulate, lobes ovate-lanceolate, acuminate, fleshy and more or less granulate, with uneven edges, longer than the tube, recurving. Infrastamineal scales oblong, fringed, scarely reaching the stamens. Anthers oval-oblong on very short filaments. Styles slender and shorter than the conic ovary; stigmas small, capitate. Capsule

globose-conic, enveloped by the withered corolla, mostly 1-seeded; embryo large, globose, with short terete appendage.

Nevada and California on Artemisia, Atriplex, Suaeda, etc.

Nevada: Clark and Nye Counties

. CUSCUTA CALIFORNICA Choisy var. APICULATA Engelmann, Trans. Acad.
   Sci. St. Louis 1:   499.   1859.

Flowers about 2.5 mm. long from base to corolla sinuses, pedicell-ate in few-flowered clusters. Calyx somewhat thickened and fleshy at the base, lobes ovate-lanceolate, acuminate, about reaching the corolla sinuses. Corolla campanulate, lobes lanceolate, acuminate, erect to spreading, longer than the tube. Infrastamineal scales lacking. Stamens shorter than the corolla lobes; anthers oblong, about equal to the slightly subulate filaments. Styles slender, commonly longer than the ovoid-pointed ovary; stigmas capitate. Capsule ovoid-pointed.

This variety was originally described from a specimen collected by Bigelow "On the Colorado." The two Nevada specimens cited below are the only other specimens which have been seen of this variety.

Nevada: Clark County; Valley of Fire, 2500 feet altitude, in fine sand, May 28, 1937, I. LaRivers & N. F. Hancock, No. 231; 4 miles southeast of Muddy Peak at head of Callville, Wash., 2500 feet altitude, sandy washes of Covillea belt, May 29, 1937, I. LaRivers & N. F. Hancock, No. 263.

8. CUSCUTA OCCIDENTALIS Millspaugh, in Millspaugh & Nuttall, Fl.

   Santa Catalina Isl., Field Mus. Nat. Hist. Bot. 5: 204. 1923.

   Cuscuta californica Choisy var. breviflora Engelmann,

   Trans. Acad. Sci. St. Louis 1: 499. 1859.

   Flowers about 2.5 mm. long from base to corolla sinuses, sessile or subsessile in compact clusters of several flowers. Calyx thickened and fleshy at the base, lobes ovate-lanceolate, acuminate, about reaching the corolla sinuses, or longer. Corolla campanulate, becoming urceolate about the maturing capsule, lobes lanceolate, acuminate, equaling or exceeding the tube in length, abruptly spreading in fruit to produce a characteristic star-shape as the flower is viewed from above. Infrastamineal scales lacking. Stamens much shorter than the corolla lobes, anthers rounded or oval. Styles slender, about as long as, or sometimes longer than, the globose ovary; stigmas small, capitate. Capsules globose, thin, enveloped by the withered corolla.

   Pacific coast states and ocasional eastward to western Colorado on a variety of hosts.

   Nevada: Mineral, Lye, ?Humboldt, and ?Clark Counties.

9. CUSCUTA EPITHYMUM Murray, in Linn. Syst. Veg. ed. 13, 140. 1774.

   Stems slender, sometimes purplish red. Flowers about 2 mm. long from pedicel to corolla sinuses, sessile in dense, glomerulate clusters. Calyx lobes triangular to sublanceolate, acute, commonly about as long as the corolla tube, sometimes tinged with purple. Corolla

lobes triangular to ovate-lanceolate, acute, spreading, mostly
shorter than the tube. Infrastamineal scales fringed about the
upper part, not reaching the stamens. Stamens shorter than the
corolla lobes, filaments longer than the oval to subsagittate anthers. Stigmas filiform, about as long as the styles. Capsules globose, circumscissile, capped by the withered corolla, seeds usually 4.

This species is common throughout Europe where it parasitizes a
great variety of hosts, though preferring legumes. It has been distributed throughout the range of the genus with the seeds of its
legume hosts. It is found occasionally in the United States. It
may cause considerable damage, especially to clover.

Nevada: Washoe County.

10. CUSCUTA APPROXIMATA Babington var. URCEOLATA (Kunze) Yuncker,
    Mem. Torr. Bot. Club 18: 297. 1932.
    Cuscuta urceolata Kunze, Flora 4: 651. 1846.
    Cuscuta planiflora Tenore var. approximata Engelmann, Trans.
        Acad. Sci. St. Louis 1: 465. 1859.

Flowers 1.5 to 2 mm. long from base to corolla sinuses, sessile
in compact, few- to several-flowered glomerules. Calyx enclosing
the corolla tube, shallowly divided, lobes mostly wider than long,
with fleshy, turgid tips, cells large and prominent and commonly
glistening golden-yellow when dry. Corolla campanulate but soon
becoming globose about the developing capsule, lobes ovate- or oval-
orbicular, spreading. Infrastamineal scales oblong, shallowly

fringed about the top, reaching the stamens or shorter. Stamens shorter than the corolla lobes. Stigmas filiform and about equal to the styles, often red; ovary depressed-globose. Capsule depressed-globose, readily circumscissile, enveloped by the withered corolla, commonly 4-seeded.

This species is widely distributed throughout Europe and western Asia where it occurs most commonly on leguminous hosts. It has been introduced and is now occasional to frequent in the western United States, often on alfalfa where it may do considerable damage.

Nevada: Ormsby and Washoe Counties.

Lightning Source UK Ltd.
Milton Keynes UK
UKHW041152150219
337137UK00013B/1571/P